Understanding Ecosystems

Printed in the United States of America

ISBN-13: 978-0-15-362040-9

ISBN-10: 0-15-362040-4

1 2 3 4 5 6 7 8 9 10 179 16 15 14 13 12 11 10 09 08 07

Visit *The Learning Site!*
www.harcourtschool.com

Lesson 1

What Are the Parts of an Ecosystem?

VOCABULARY

environment
ecosystem
population
community

An **environment** is all of the living and nonliving things that affect an organism.

An **ecosystem** is all the living and nonliving things in one place. An ecosystem can be as small as the space under a rock or as large as a forest.

A **population** is a group of all of the same kind of living things living in the same ecosystem. These water lilies are part of one population.

A **community** is made up of all of the populations of living things that live in the same place.

READING FOCUS SKILL
MAIN IDEA AND DETAILS

The **main idea** is what the text is mostly about. **Details** tell more about the main idea.

Look for **details** that tell about an ecosystem.

Ecosystems

An **environment** is all the living and nonliving things that affect a living thing. The living things in your environment are people, animals, and plants. The nonliving things are water, air, soil, and weather.

An **ecosystem** is all the living and nonliving things in a place. It also has a climate. Climate is the average weather over many years. It includes temperature and rainfall.

▼ Prairie smoke plants

▼ Prairie dogs

Ecosystems may be different sizes. A small ecosystem might be the space under a rock. Insects may live there. Nonliving things include soil, water, and air.

A large ecosystem might be a forest. This ecosystem is home to many plants and animals. Each living thing finds what it needs in the forest. A forest also has nonliving parts. They include water, air, soil, and climate.

Focus Skill **Name the two parts of every ecosystem and give two examples of each part.**

Moose live in a forest ecosystem. ▶

Individuals and Populations

One plant is an individual. One animal is an individual, too. A bee or pine tree is an individual.

A **population** is a group of the same kind of individuals living in the same ecosystem. Examples of populations include a hive of bees or a group of pine trees. Members of some populations, like frogs, live alone. Other populations, like wolves, live in groups.

Blackbird population ▼

Some populations can live in different places. For example, red-winged blackbirds live in wetlands. But they also live in other areas. Because they can live in different areas, a population of red-winged blackbirds could move to a different area if their home no longer met their needs.

Some populations can only live in one area. These organisms cannot move if their home no longer meets their needs. This means the population if usually small.

Name another example of an individual and a population.

Communities

A **community** is all the populations that live in the same place. It includes all the plants and animals that live there.

The plants and animals in a community depend on each other. Some animals eat the plants. Other animals eat the plant eaters. Animals help plants by spreading seeds. Plants provide places for animals to live and hide.

Many populations make up the communities in this cold taiga ecosystem. They include conifers, moose, and many kinds of birds.

An example of a community is the area in the Everglades National Park. There are many different populations that live there. There are alligators and raccoons. There are 350 different kinds of land birds. There are 16 kinds of wading birds. There are many different kinds of plants.

All of the organisms in the Everglades National Park depend on one another. The raccoons help by spreading seeds. The fish are food for the alligators. The whole community would be in danger if you took an organism out of it.

How do plants and animals in a community depend on each other?

Review

Focus Skill

Complete this main idea statement.

1. A forest _______ is made up of both living and nonliving things.

Complete these detail statements.

2. A rabbit is an example of an _______.
3. A group of robins living in a forest is called a _______.
4. A _______ includes all the plants and animals that live in the same place.

Lesson 2

What Factors Influence Ecosystems?

VOCABULARY
biotic
abiotic
diversity

The living parts of an ecosystem are **biotic**. Plants and animals are biotic factors in an ecosystem. The nonliving parts of an ecosystem are **abiotic**. Air, water, sunlight, and soil are abiotic factors in an ecosystem.

Diversity is the variety of living things. A rain forest has the most diversity of all ecosystems.

READING FOCUS SKILL
CAUSE AND EFFECT

A cause is what makes something happen. An effect is what happens.

Look for the effects that living and nonliving things have on ecosystems.

Living Things Affect Ecosystems

Plants and animals are biotic factors in an ecosystem. **Biotic** factors are the living parts of an ecosystem. These factors affect the ecosystem. They also affect each other in many ways.

Animals use plants in many ways. Plants provide food and shelter.

Animals can help plants. By eating plants, they give more space for other plants.

Animals can also harm plants. Too many plant eaters can wipe out plants.

Gypsy moths stripped these trees. ▼

Animals can affect one another. More wolves can lead to more rabbits being eaten. Fewer rabbits may cause hungry wolves. Fewer rabbits may cause more plants to grow.

A change in plants can cause a change in animals. If dry weather kills plants, rabbits have less to eat. Fewer rabbits causes wolves to go hungry.

Sometimes new plants or animals can change an ecosystem. New plants can crowd out other plants. New animals may wipe out other animals. Or they may eat all the plants.

Disease can change an ecosystem. It can kill plants or animals. This can change an ecosystem's balance.

Focus Skill **Tell how an increase in plants could affect an ecosystem.**

Too many deer can wipe out plants. ▶

Nonliving Things Affect Ecosystems

Air, water, soil, and sunlight are the abiotic parts of an ecosystem. **Abiotic** factors are the nonliving parts of an ecosystem. They are as important as the living things in an ecosystem.

A change in water can affect all living things in an ecosystem. Too little rain causes plants to die. Without plants, animals may die or leave.

Soil also affects an ecosystem. Rich soil means many plants will grow. Poor soil leads to fewer plants. Without plants, fewer animals can live in the ecosystem.

Air also affects living things. Bad air can harm many living things.

Tell how a change in the water supply could affect a rabbit.

Nonliving factors ▶

sunlight
air
water
soil

Climate Affects Ecosystems

Climate is an abiotic factor. Climate includes the amount of rainfall and sunlight. It also includes air temperature. This map shows the main climates of Earth.

Climate affects the soil. Some climates allow many living things to grow.

Climate also affects the kinds of living things in an ecosystem. Some living things need mild or wet climates. Others are suited to cold or dry climates.

Tell what would happen to an ecosystem if its climate changed.

Main climate zones of Earth ▼

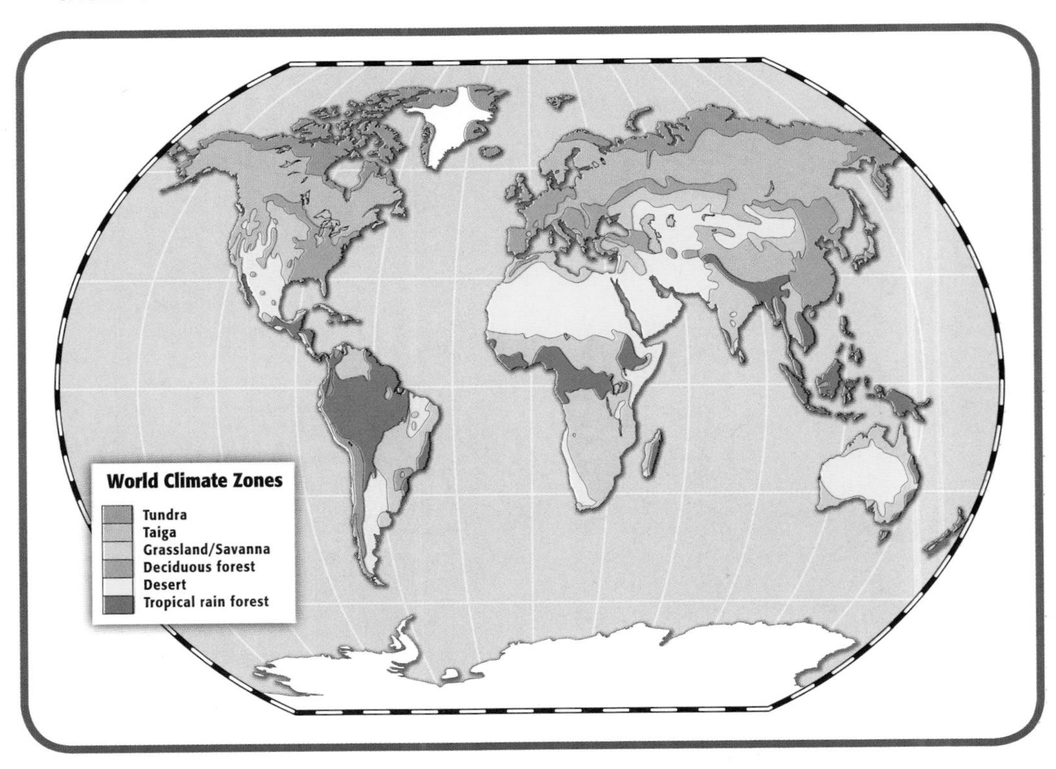

Diversity

Some ecosystems have only a few living things. Others have many. **Diversity** is the variety of living things.

Rain forests have the most diversity of all Earth's ecosystems. This is because they provide the things that living things need.

Other ecosystems have little diversity. In these ecosystems, living things may find it hard to survive.

Tell what leads to a diversity of living things in an ecosystem.

Rain forest ecosystem ▶

Review

Focus Skill

Complete these cause and effect statements.

1. A decrease in wolves can cause an ________ in rabbits.
2. Poor soil may lead to ________ plants.
3. Too little rain can cause plants and animals to ________.
4. The ________ of plants and animals in an ecosystem is a result of how many living things can find what they need to live.

Lesson 3

How Do Humans Affect Ecosystems?

VOCABULARY

pollution
habitat
restoration

Harmful waste products that enter an ecosystem are **pollution**. Pollution may harm water, land, or air.

Habitat restoration happens when people return a natural environment to its original condition. Planting trees is one way to repair damage to an environment.

READING FOCUS SKILL
COMPARE AND CONTRAST

When you **compare and contrast**, you tell how things are alike and different.

Compare positive and negative effects of changes made by humans on ecosystems.

Humans Within Ecosystems

Natural resources are parts of an ecosystem that people use. They include air, water, soil, trees, and minerals.

Some resources, such as air, water, and trees, can be replaced. Others cannot. Resources that cannot be replaced include coal, oil, and iron. Once supplies are used, these resources will be gone.

Lakes, fresh air, and sunlight are natural resources. ▼

People use resources in many ways. They use wood to build homes. They use sand to make glass. They use iron to make steel.

People raise crops to feed themselves. They also raise crops to make medicines.

How are coal and water alike and different?

▲ Medicine is made from this tree.

Wheat is used to make bread. ▼

Negative and Positive Changes

People's actions can harm an ecosystem. People may clear land. This can destroy habitats. Animals can no longer meet their basic needs. They must move or die.

People can cause pollution. **Pollution** happens when harmful materials mix with water, air, or soil.

Water pollution can happen if chemicals wash into water. Trash can also pollute water. Soil pollution can come from trash and chemicals.

Air pollution can come from cars. Smoke from smokestacks can also pollute the air. Some chemicals form acid rain. This harms plants and pollutes lakes.

◀ **Cars can cause air pollution.**

Tree planting ▶

People help ecosystems by habitat restoration. **Habitat restoration** happens when people return a natural environment to the way it was. Planting trees can do this. Making wetlands can, too.

People also help when they pollute less. They do this by reducing harmful gases. Recycling and using fewer chemicals also helps ecosystems.

Tell how people can harm and help an ecosystem.

▼ Bicycles don't release pollution.

Protecting Our Air and Water

Living things need air and water. Clean air and clean water are important. Governments, businesses, and communities can act to make sure we have clean air and water. So can you!

Laws have been passed to make sure companies and factories do not pollute the air. You can help stop air pollution, too. When you save electricity, you cut down on how much fuel has to be burned to produce energy. Turn out the lights when you are not in a room. Reuse grocery bags. Walk or ride your bike or take the bus instead of driving a car when you are making short trips.

▼ **Factories have had to install scrubbers to clean the smoke they release into the air.**

Before

After

▲ **The blue water might look clean but it is full of chemicals. People cleaned the water up and now the ecosystem is healthier.**

Protecting our water is also important. Laws have been passed to make sure water pollution is prevented.

You can take short showers to save water. You can turn off the tap when you are brushing your teeth. There are many things you can do to help protect our air and water!

What are some ways to protect the air?

Planning for Change

Earth's human population keeps growing. More people need more homes, roads, and factories. Building can harm ecosystems.

Before building, people must plan. They must think about biotic factors in an ecosystem. Filling wetlands can harm plants and animals. Washing soil into a river can harm fish and plants.

Planners draw plans and build models. ▼

Planners must also think about abiotic factors. This includes climate, rainfall, and type of soil.

Careful planning helps protect the community and physical environment of an ecosystem.

Tell about the different kinds of factors people must think about when building.

Construction site ▲

Review

Complete these compare and contrast statements.

1. Oil, wheat, air, and iron are all _______.
2. Trees, water, and air are resources that can be _______, while oil and coal cannot.
3. Both recycling and reducing the use of chemicals can help to reduce _______.
4. Planting trees and creating wetlands are both examples of _______.

GLOSSARY

abiotic (ay•by•AHT•ik) of the nonliving parts of an ecosystem

biotic (by•AHT•ik) of the living parts of an ecosystem

community (kuh•MYOO•nuh•tee) all the populations of organisms living together in an environment

diversity (duh•VER•suh•tee) a great variety of living things

ecosystem (EE•koh•sis•tuhm) all the living and nonliving things in one place

environment (en•VY•ruhn•muhnt) all of the living and nonliving things that affect an organism

habitat restoration (HAB•i•tat res•tuh•RAY•shuhn) returning a natural environment to its original condition

pollution (puh•LOO•shuhn) waste products that damage an ecosystem

population (pahp•yuh•LAY•shuhn) all the individuals of the same kind living in the same environment